Interacting Life: The Ecosystem

Copyright © by Harcourt, Inc.

All rights reserved. No part of this publication may be reproduced or transmitted in any form or by any means, electronic or mechanical, including photocopy, recording, or any information storage and retrieval system, without permission in writing from the publisher.

Requests for permission to make copies of any part of the work should be addressed to School Permissions and Copyrights, Harcourt, Inc., 6277 Sea Harbor Drive, Orlando, Florida 32887-6777. Fax: 407-345-2418.

HARCOURT and the Harcourt Logo are trademarks of Harcourt, Inc., registered in the United States of America and/or other jurisdictions.

Printed in Mexico

ISBN 978-0-15-362264-9
ISBN 0-15-362264-4

2 3 4 5 6 7 8 9 10   050   16 15 14 13 12 11 10 09 08

Visit *The Learning Site!*
www.harcourtschool.com

Energy Exchange

What's one thing that you need to work, play, and even sleep? You need energy. You get energy from the food you eat. In fact, every living thing needs food in order to produce energy.

Energy begins with the sun. Plants absorb the sun's energy and use it to take in nutrients from the soil, carbon dioxide from the air, and water. They also use the energy from the sun to turn those elements into a type of sugar called glucose. Some of the glucose is stored in the plant as energy. The stored energy is passed along when an animal or person eats the plant.

Plants make their own food using a process called *photosynthesis*.

Caterpillars are consumers that eat a large amount of food for their weight. One caterpillar can eat twice its weight every day.

Plants are producers within any ecosystem. A **producer** is an organism that makes its own food. Any organism that eats another organism is called a **consumer**. Bushes, grass, and trees are all producers. Deer that eat leaves are consumers. A wolf that eats the deer is also a consumer.

Producers and consumers form parts of a food chain. A **food chain** is the line of producers and consumers that pass energy from one to another. A food chain might begin with green grass that makes its own food from the sun's energy. In a forest ecosystem, a rabbit that eats the grass will get some of the stored energy. The rabbit will use some of the energy to stay alive and will store the rest within its body. A fox that eats the rabbit will receive the rabbit's stored energy. You can see how the original energy is passed from one living thing to another.

 SEQUENCE Give an example of three organisms that form a food chain.

Eat or Be Eaten

The fox in a forest ecosystem is a predator. A predator is an animal that eats other living animals. The rabbit is the fox's prey. Prey are the animals eaten by other animals. The energy flow through the forest ecosystem doesn't stop with the fox. The fox isn't the top predator in the forest. It, too, is prey at certain times. Larger predators such as wolves will hunt and eat foxes.

Another important role in an ecosystem is that of a scavenger. A scavenger is an animal that eats other animals that have died. Crows are scavengers within a forest ecosystem.

Fast Fact

One type of decomposer, the Japanese fungus, produces enzymes that make it glow in the dark. Its glow can be seen from 15 m (50 ft.) away!

Lions are the top predators on the savanna. They have no natural enemies other than humans.

More than half of the air in our atmosphere is made up of nitrogen gas.

All living things are made up of cells. Cells are made up of proteins, which in turn are largely made up of nitrogen. Living things can't just get nitrogen from the air the way they get oxygen. That's where other important organisms of an ecosystem come in. These organisms are decomposers such as mushrooms and other fungi, and bacteria. Decomposers break down animal and plant remains. They convert the nitrogen back into substances that can be used again by living things to make proteins. This process gives decomposers their food. It also returns nutrients—including nitrogen—back into the soil. Plants grow in the nutrient-rich soil to complete the circle. Animals that eat the plants will receive the nitrogen and other nutrients they need.

SEQUENCE Explain the role of predator, prey, and scavenger within an ecosystem.

Food Webs

Organisms in a food chain often eat—and get eaten—by more than just one thing. Rabbits don't eat just grass. They also eat clover, leaves, and shrubs. In turn, the grass isn't eaten just by rabbits. Deer and small mammals also eat grass. Rabbits aren't eaten just by foxes. They are also prey to coyotes and wolves. Foxes eat squirrels, toads, chipmunks, and rabbits. This connection among food chains forms a food web. **Food webs** are overlapping food chains within an ecosystem.

Food chains and food webs occur in the water as well as on land. Even in the water, almost all food chains begin with energy from the sun. Tiny organisms in the ocean called phytoplankton make their own food from the sun's energy. They also put oxygen back into the air and serve as a food source for many other creatures.

Fast Fact

Blue whales, the largest animals on earth, feed on tiny krill. They can eat 4 tons (3.6 metric tons) of krill in a day.

Like those on land, food webs in the ocean are made up of overlapping food chains. Very small animals called krill eat phytoplankton. Squid and tiny fish eat krill. Larger fish eat those fish and squid. Seals eat these fish, and then the seals are eaten by sharks or other larger predators. Each animal receives some stored energy from the animal or plant that it eats.

Freshwater lakes and ponds have their food webs, too. Small plants and algae grow and make their food from the sun's energy. Small fish eat the plants. Larger fish eat the smaller fish. These are then eaten by mammals such as otters and large wading birds such as storks. Alligators are top predators that eat birds, fish, and small water mammals.

 COMPARE AND CONTRAST Explain the differences and similarities between a food chain and a food web.

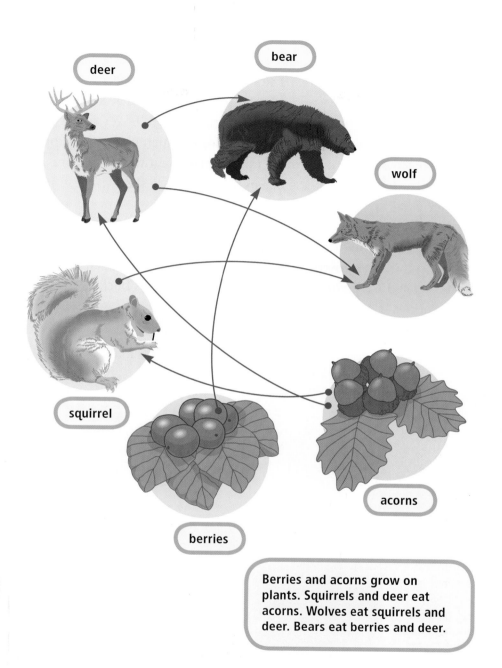

Berries and acorns grow on plants. Squirrels and deer eat acorns. Wolves eat squirrels and deer. Bears eat berries and deer.

Energy and Cycles in Ecosystems

Some energy is lost every time it moves through to the next step of a food chain. When a deer eats leaves, it is able to use only about ten percent of the energy that was stored in the leaves. When a wolf eats the deer, it is able to use about ten percent of the energy stored in the deer. As you go higher up in a food chain, there are fewer and fewer animals near the top, because they require so much energy to support them. That's why there are more rabbits in a forest than there are foxes, or more tiny fish in the ocean than there are sharks. These stages of energy gain and loss form an energy pyramid.

Energy is constantly recycled back into the environment through food webs and other cycles.

Using the sun's energy, plants put water back into the air through a process called *transpiration*. Eventually, the water falls back to the earth as rain and is used again by all living things.

This consumer is able to use only about ten percent of the energy stored in the plant it eats.

The carbon cycle is another important part in the function of ecosystems. Animals breathe in oxygen and breathe out carbon dioxide. Plants use the carbon dioxide to make food. Plants then release oxygen back into the air for use by animals.

Carbon can also be cycled by decomposers. It can also become stored in fossil fuels. When fossil fuels are burned they release carbon back into the environment.

The nitrogen cycle is a way that nitrogen is cycled through the environment for use by various organisms. Decomposers are an important part of this cycle.

 MAIN IDEA AND DETAILS Explain the importance of carbon dioxide for both animals and plants.

In photosynthesis, plants release oxygen that animals need to stay alive. This process is part of the carbon cycle.

Competition

What happens when creatures in the wild compete for the same things? **Competition** is the struggle among different organisms for limited resources such as food, water, space, light, and even air. For example, there might not be enough mice one year to feed all the owls in one ecosystem, so all the owls in the area compete for the few mice. Some owls will go hungry. Others won't survive or be strong enough to have young, and the owl population will decrease. Because there are fewer owls eating mice, the mouse population will increase. Once the population increases, there will be more food for the owls. Then the owl population will increase because there will be plenty of food. This is a good example of how a change in one population of organisms can greatly affect another.

Pandas eat only bamboo. As bamboo forests are destroyed, pandas compete for fewer and fewer food sources. That's one reason pandas are endangered.

When people kill wolves, the wolves' prey increases because wolves aren't around to keep populations down. The entire ecosystem becomes unbalanced.

Because a food web is made up of interconnecting food chains, many organisms can be affected when one member of an ecosystem is out of balance. For example, birds that feed on only one type of tree can't survive when that type of tree is cut down. The larger mammals that eat those birds can't find food and begin to die off too. The insects that the birds eat will increase in numbers out of control because they are living without their natural enemies to consume them. Too many insects kill off plants. This leads to less food and shelter for other animals, too much sunlight, and not enough oxygen released back into the air. Just one change in a food web can affect an entire ecosystem because organisms are so connected.

Fast Fact

The white-tipped sicklebill feeds only on heliconia flowers because its long, hooked-shaped beak can only fit inside these flowers.

CAUSE AND EFFECT Give one example of how the decrease and increase in one population of an animal or plant affects the population of another within the same ecosystem.

Working Together

Different species within ecosystems work together in other ways, too. Some have an association called *symbiosis*. **Symbiosis** is a close relationship between organisms of different species in which one or both benefit. There are three different kinds of symbiosis.

One is called *parasitism*. An organism that benefits from its relationship with another organism while the other organism is harmed is called the **parasite**. An example of this is fleas living on dogs. In this case a flea is the parasite. The animal on which the parasite lives, such as the dog, is called the **host**. The fleas get food by drinking the dog's blood. However, the dog is harmed. The dog becomes very uncomfortable and it itches and scratches. In extreme cases, the dog can die from blood loss due to the fleas.

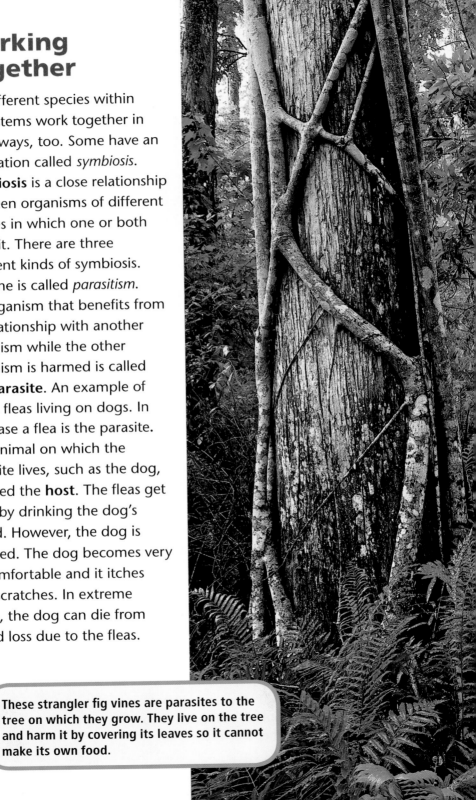

These strangler fig vines are parasites to the tree on which they grow. They live on the tree and harm it by covering its leaves so it cannot make its own food.

Another type of symbiosis is called mutualism. This occurs when both organisms benefit from their relationship.

For example small white birds called oxpeckers sit on the backs of rhinoceroses. The birds pick and eat little insects on the rhino's back. The rhinos have poor eyesight and actually benefit by their relationship with the birds. When the birds see danger, they fly away, which warns the rhinos to be aware and on alert for danger. In addition, the birds remove irritating insect pests.

Fast Fact

Jays often encourage ants to crawl all over their feathers. The ants produce an acid that kills the parasites that make their home in the feathers.

The third type of symbiosis is called commensalism. Commensalism is a relationship in which one species benefits and the other isn't harmed or helped. Orchids in the rainforest benefit by growing high up on the branches of other trees. By being this high they receive a lot of light, moisture, and nutrients. The trees who host them aren't helped by the orchids, nor are they hurt by them.

 COMPARE AND CONTRAST How are parasitism and commensalism alike? How are they different?

This is an example of commensalism. The bird benefits by using a tree to build its nest, and the tree is neither harmed nor helped in the process.

A Delicate Balance

On the surface, the interruption of a link in a food chain or web might not seem like it would have much of a consequence. But the consequences are huge. Land development, hunting, and air and water pollution affect many living organisms. They can lead to a reduction of sunlight, territory, water, and air, and in turn reduce the numbers of consumers and producers.

A change in one symbiotic relationship can greatly affect an entire ecosystem. Predator and prey relationships, symbiosis, and the different jobs of decomposers and scavengers all play an important role in keeping nature's delicate balance alive and well.

Fast Fact

One tiger needs a lot of jungle territory to hunt its prey. As these lands are destroyed, tigers' territories overlap. Then they compete for less food and their populations decline. Scientists are now using hidden cameras to study tiger populations.

Tropical rain forests around the world are ecosystems that depend on an extremely delicate balance. Upsetting this balance can greatly affect plant and animal life.

Pesticides not only kill bugs, they can hurt the animals that eat the bugs, such as this bat. They also hurt the animals that eat bats!

 MAIN IDEA AND DETAILS Explain the importance of plants to the survival of other species.

Summary

Every living thing needs energy. The sun is the ultimate source of all energy in an ecosystem. Sunlight is needed for plants to grow. Plants serve as food for other organisms as well as a home for them. Plants can return some nutrients to the soil. Plants absorb carbon dioxide from the air and give back oxygen and water. Animals breathe the oxygen and feed off the plants. Larger animals feed off smaller animals that eat the plants. All of these things play an important role in food chains and food webs. That's why even one disturbance can be a problem within an ecosystem. For example, if the trees on which orchids grow are destroyed, these flowers can't survive. The insects that feed on these flowers won't survive and the animals that eat those insects won't survive. Living organisms form food chains and food webs that depend on each and every member. Everything in nature relies on other living organisms for its own survival.

Glossary

competition (kahm•puh•TISH•uhn) The struggle among organisms for limited resources in an area (10)

consumer (kuhn•SOOM•er) An organism that eats other organisms (3, 14)

food chain (FOOD CHAYN) A sequence of connected producers and consumers (3, 6, 7, 8, 11, 14, 15)

food web (FOOD WEB) A group of connected food chains in an ecosystem (6, 7, 9, 11, 15)

host (HOHST) The organism that a parasite lives in or on (12, 13)

parasite (PA•ruh•syt) An organism that benefits from its relationship with another organism while the other organism is harmed (12, 13)

producer (pruh•DOOS•er) An organism that makes its own food (3, 14)

symbiosis (sim•by•OH•sis) A close relationship between organisms of different species in which one or both of the organisms benefit (12, 13, 14)